The Full-Blown Denier

D.L. Bjorngjeld

Published by D. L. Bjorngjeld
Forest Lake, Minnesota

ISBN 978-0-9981110-6-3

Layout & Design by Todd Anderson

Printed in United States of America

Part One

170 Years of Warming

Man-Made

Or

Natural Forces

Chapter One

"So, you're a serious denier these days," an old friend from school – at least I thought he was still a friend, though I hadn't seen him in a while – said with a snarky sort of tone.

"Denier of what?" I asked, chuckling.

"Global warming... I saw the books you wrote."

I paused for a moment, still smiling, then said, "If you mean our climate has been slowly warming for a hundred and seventy years or so, I don't deny that. BUT, if you are referring to what the Alarmists have been preaching for the last three decades, then YES, I am a Full-Blown Denier."

"So, you know more than 97% of all the scientists," he said, again with the snarky tone.

I scratched my forehead, then said, "I'm guessing that you haven't actually read my books. What is it that you think 97% of all scientists agree on about climate?"

"Burning fossil fuels and causing large CO_2 emissions is what's causing the dangerous warming."

"And where did you learn that?" I asked.

"From the Consensus," he insisted, sounding as though he was flabbergasted and thought, *How could you be such a fool and not know this.*

"Do you know where the '97% Consensus' idea came from?" I asked.

"Not really. I suppose from surveys of scientists. Maybe from the United Nations."

"Well, sort of," I said. "One of the first such 'surveys' was done by a University of Illinois student, Kendall Zimmerman, and her advisor, Peter Doran. They emailed a two question survey to more than 10,000 members of the American Geophysical Union." (I quickly found the following graph in my book and showed it to him as we spoke.)

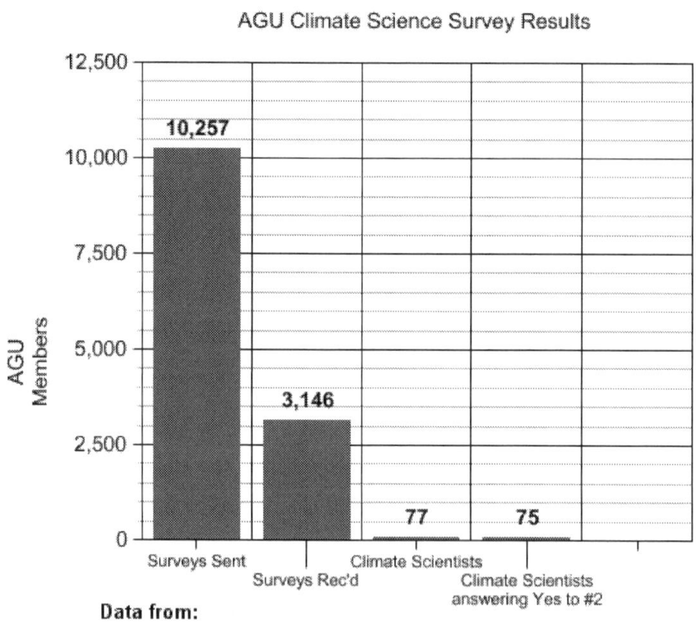

Data from:
Doran, P. T. and M. K. Zimmerman (2009), Examining the Scientific Consensus on Climate Change, Eos Trans. AGU, 90(3), 22, doi: 10.1029/2009EO030002.

"They received just over 3,100 responses. I think all of those who responded said Yes to: 'Statement #1 – Global temperatures have risen since 1900.'

“From the 3,100 responses that were received, they hand-picked 77. Of the 77 selected, 75 said Yes to: ‘Statement #2 - Human activity is the significant contributing factor to global warming.’

“So, they have 75 scientists out of 77... Voila... 97%. That’s *SEVENTY-FIVE* scientists from the AGU, out of more than 10,000 members originally surveyed, and from more than 3,100 that responded to the survey, and that’s what they call a ‘*97% CONSENSUS*.’ There’s been several other similar surveys, all with the same result, all based on similar questionable data.”

“That’s just denier propaganda,” he said.

“I tell you what... I’ll challenge you to a $10 wager,” I said. “Just a fun bet on who can come up with the best answer as to what the 97% Consensus is based on.” I reached out my hand, waiting for his response.

"Okay," he finally said, and shook my hand. "But it's not quite a fair bet, because you know from our school days how I hate doing that kind of research."

"Yes, I do know that. ***And so does the IPCC and their loyal ‘Acolytes’*** who all feed at the trough of huge Government Grants. ***They count on the fact*** that the general public won't bother questioning or researching anything they say. They know that a good share of regular folks will hear every statement, every scare tactic that they throw out there and swallow them hook, line, and sinker. Never questioning or digging into any of it.”

He thought for a moment, then asked, “What’s our deadline? When do we get together and compare notes?”

“How about we meet on the first of next month for dinner? That gives us almost three weeks,” I said.

I jotted down a few web sites he could check out, along with Google and Wikipedia. “When it comes to the topic of Global Warming, I don’t trust what’s on Wikipedia any farther than I could throw an elephant.”

“Why’s that?” he asked, a little more friendliness and curiosity in his tone.

“In the early 2000s, a fellow named William Connolley was given full authority by Wikipedia to delete or rewrite any article relating to global warming, and to ban anyone he disagreed with. In the next few years he banned nearly 2,000 people, and deleted or rewrote 5,400 articles. So, virtually everything on the site is slanted in favor of the Alarmist's argument. You can even check-out the Connolley thing on Google. Just enter *William Connolley – Wikipedia.* It’s kind of amazing that they even put that stuff about Connolley on their site,” I added. “It’s like saying, ‘You can trust us for information on Global Warming, even though we allowed this guy to delete or rewrite every article pertaining to the subject.’”

That encounter with my old friend – along with talking to a couple of other friends who had read my books and weren’t exactly sure of the details – prompted me to write this, "The Full-Blown Denier.” With any luck, it will be a blending of "A Brief Look at Earth's Beautiful, Chaotic Climate" and "Grand Solar Minimum." I’m trying for something clear and easily understood. And, please, if there’s anything you disagree with show me where you think I’m wrong. By that I mean show me evidence, not statements of “opinion.” The IPCC regularly uses statements like “***Highly likely***”... “***Compelling***”... “***High confidence level***”... Those are statements of opinion, not of fact. Virtually everything put out by the IPCC (and by the web sites supporting their theory) is loaded with those kinds of qualifying statements of opinion, not of scientific, provable evidence.

Chapter Two

My book, "A Brief Look at Earth's Beautiful, Chaotic Climate," was meant to be one person's opinion (my opinion) of the illogical belief – or dogma, since it comes very close to being a religious conviction for many followers – professed by Alarmists and their Disciples. The book was not meant to be a scientific or expert opinion, but that of a reasonable and logical person who has studied everything available on Global Warming for more than fifteen years.

Early on, I saw a pattern that was centered on the idea of "Consensus." When all else fails, ***"The science is settled and supported by the Consensus"*** is their fallback position. Science is not proven nor disproven by consensus. As Albert Einstein put it, "A hundred experimentations can try to prove me right, but it only takes one to prove me wrong." Or Thomas Huxley, who said the great bane of science is, "A beautiful hypothesis destroyed by an ugly fact." I repeat, science is not proven nor disproven by a consensus. In fact, whenever you hear the term "consensus" used in relation to science, you should immediately be suspect of a lack of facts or proof supporting their theory (and carefully guard your wallet).

The second pattern I saw was the need for an immense, dramatic "The Sky is Falling" mentality. There is a continual need to stir fear in the minds of the public, the vast majority of whom, they know, will not question, doubt, or research whatever the latest ploy is in their ceaseless stream of fear-mongering tactics.

Stephan Schneider, one of the leading Professors of Environmental Biology, and strong IPCC supporter, wrote, ***"To reduce the risk of losing our grip on 'Disastrous Climate Change,' we need to capture the public's attention and imagination. So, we have to offer up scary scenarios, make dramatic statements, make no mention of our doubts, and find the right balance between being honest and being effective."*** (Bold italics and underlining added.)

Michael Hulme, Professor of Climate at King's College London, and a former Professor of Climate Change at the University of East Anglia in Norwich, England wrote, ***"We will continue to create, write and tell new and alarming stories about Climate Change. And we'll insist that fellow Alarmists keep beating the drum and telling the stories."***

"Offering up scary scenarios..." "Making dramatic statements..." "Creating new and alarming stories about climate..." "Finding the right balance between being honest and being effective..." These are tactics that you use when you don't have the facts and evidence on your side.

Patrick Moore, Ecologist, Environmentalist, and 1970 co-founder of Greenpeace insists, "Man-Made Global Warming is **a theory for which there is no scientific proof what-so-ever.** There is only a large, vigorous, vocal group on the left that are ferociously evangelizing."

Climatologist, Dr. Timothy Ball stated, "**There is no record, no evidence, no proof anywhere in climate history which shows an increase in CO2 will cause temperature to rise more than minimally.** This is the Alarmists fundamental theory, while all evidence shows that the opposite is true, **that warming temperatures cause atmospheric CO2 levels to rise.**"

This brings up another significant item that Alarmists never divulge... That the CO_2 emissions caused by human activities constitute only 3 to 4% of all CO_2 released into the atmosphere every year. The overwhelming majority, 96-97%, is released from natural sources (the oceans, volcanoes, and others). In every 85,000 molecules of our atmosphere, 1 molecule is CO_2 caused by humans, 32 are CO_2 from natural sources. The Alarmists want to convince you that our 1 molecule is the cause of the warming. (You can just ignore the 32 molecules coming from natural sources.)

So, the Alarmists start with a ***fundamental theory*** that is ***fundamentally wrong***. Any increase of CO_2 in the atmosphere has a minuscule influence on temperature.

Over Earth's 4.5 Billion years, there have been times of extreme cold, while having very high levels of CO_2 in the atmosphere. The Ordovician Ice Age is one example. About 438 Million years ago, Earth's temperature plunged 10°C in a relatively short time and brought about another Ice Age. Meanwhile, the level of CO_2 in the atmosphere at that time was nearly 5,000 ppm (parts per million), or 12 times today's level of 400 ppm. And still the Alarmists try to proclaim that human caused CO_2 (the 1 molecule in every 85,000) is the "Control Knob" for Earth's climate.

So, this question continues to baffle me: ***How does the IPCC start with a false premise, which is not supported by scientific evidence (and in fact is contradicted by historical scientific evidence), yet they are left to promote their flawed theory trying to convince governments and government policy makers of impending doom?*** The impending doom they've been trumpeting for three decades now, while we have yet to experience even the slightest whiff of anything "catastrophic" in our climate.

Their argument being fundamentally flawed is one reason they needed to "brand" anyone who dared disagreed with them, thus the label "denier." The label carries with it all the connotations of a "Holocaust Denier" or a person who denies the Nazi extermination of more than 6 million Jews during the Second World War. How could anyone sink so low and so evil?

The Alarmist's Dictionary:

Denier (noun): a loathsome, revolting person who refutes the truth of Man-Made Global Warming.

Chapter Three

The recent warming we've been experiencing started in the mid-nineteenth century (roughly 1850-60), when we were coming out of the Little Ice Age. Most scientists say we are still in the last stages of the Pleistocene Ice Age, which started about 2 ½ million years ago.

Technically, we are in the interglacial period called the Holocene. Interglacial Periods generally last a few thousand years with climate that is warmer and more hospitable for humans. Glacial Periods, on the other hand, last tens of thousands of years, with temperatures that are colder and with glaciers covering more of the planet. What most folks don't know – and aren't told by the Alarmists – is that the warming we've been experiencing since 1850 is bringing us out of one of the coldest periods in Earth's 4.5 billion year history, AND out of our most recent Ice Age.

In the graph on Page 10, the present time is to the far right. The lines on the graph represent Earth's Temperature and Atmospheric CO2 Levels. The line for temperature is the slightly darker shaded line (on top for the right two-thirds). As you can see, there have only been a couple of times in the last 600 million years when it's been as cold as it is now. Many scientists (A Consensus?) say the current warming is the result of the natural cycles of the Earth and the Sun, not from the flawed theory of "Man-Made Global Warming."

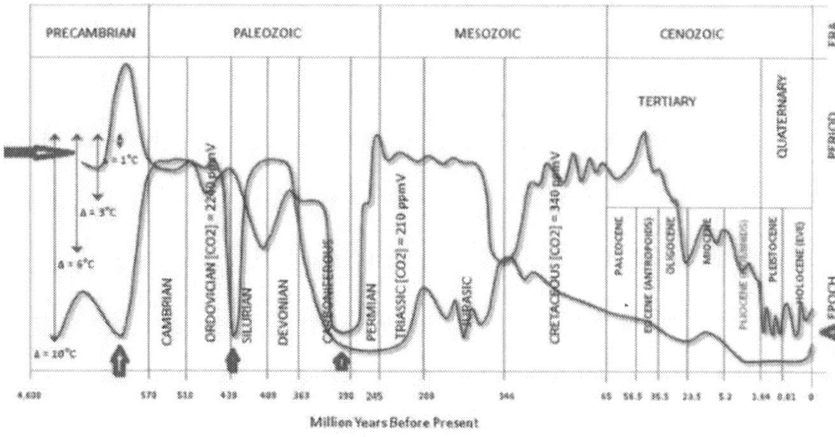

Earth's average Temperature and Atmospheric CO2 levels
for the last 600,000,000 years.

Also, as previously mentioned, if you look to the left side of the graph about 438 million years ago, you'll see the cold temperatures of the Ordovician Ice Age, while the atmospheric CO2 was nearly 5,000 ppm.

With about 12 times today's CO2 level of 400 ppm, there should have been "extreme, run-away, disastrous warming" according to the Alarmists (not falling into an another Ice Age).

In our current status – coming out of the Little Ice Age and one of the coldest periods in Earth's history – what do the Alarmists really think should be happening? That we stay in the deep freeze for another century and not have any warming? Maybe they would prefer that we sink deeper into the cold until we're in another Ice Age? I'll cast my vote for a little more warming (since they tell us they have the "Control Knob" in their hands).

Another significant factor about our climate (that is undisclosed by the Alarmists) is that for 80% of time Earth has been warmer than it is now, and with no ice anywhere on the planet. For only 20% of time has it been as cold as now, with polar ice and glaciers. From this you could conclude that the "normal" condition for Earth is warmer than now and with no ice. So, again, the fear mongering that, "We are destroying the natural ecology of Earth by causing glacial and polar ice to melt." Could it be that the "natural ecology" of Earth is warmer with no ice. The long-term view of a few billion years (I'm always amazed to think how long a Billion years is?) supports that theory. While a short-term view of 140 years might prop-up the Alarmist's theory a bit. But a period of 140 years isn't even the blink-of-an-eye in terms of geologic time or global climate considerations.

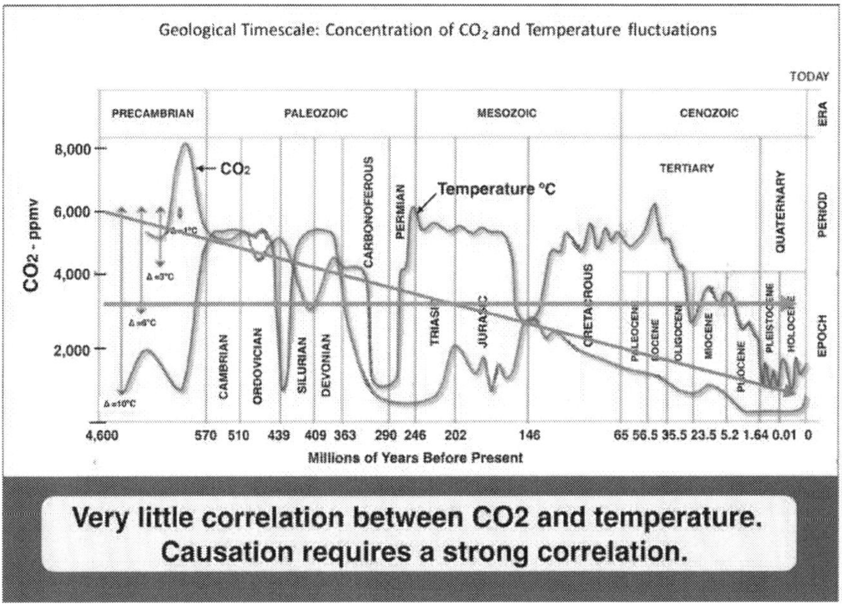

Very little correlation between CO2 and temperature. Causation requires a strong correlation.

The above graph highlights the trends of Temperature and CO2 during the past 600 million years and that there is no correlation between the level of CO2 in the atmosphere and global temperature.

"Sequestering" of CO_2 out of the atmosphere is the absorption of it into the development of limey rocks such as limestone, dolomite, granite and others. That and the evolution of vegetation, plant life, and trees that used CO_2 for photosynthesis, all greatly reduced the amount of CO_2 in the atmosphere – over hundreds of millions of years.

Fossils found in the Arctic and the Antarctic show that a variety of plant life and trees – moss, ferns, pine trees and even Arctic and Antarctic forests – thrived in those regions. And various types and sizes of animals were prevalent in those areas at different times in our history. Fossils of predators like Hyenas indicate there were a variety of prey available for hunting and scavenging, like Caribou and Mammoth. Even fossils of camel skeletons have been found in the Artic region.

Chapter Four

"This week... This Month... This year... ***Has been the warmest ever,"*** say the Alarmists at every opportunity. What they fail to mention is that it's the warmest since about 1880, when global temperature records were first being kept. (Even that is questionable, since historical records before "necessary revisions" were made showed that the warmest years were in the 1920s and 1930s.) See the following charts.

The first, the Unrevised chart, indicates several recorded temperatures that were warmer than the current "warmest ever" – one in the 1950s, one in the 1920s, and two in the 1930s, including the warmest ever in about 1934.

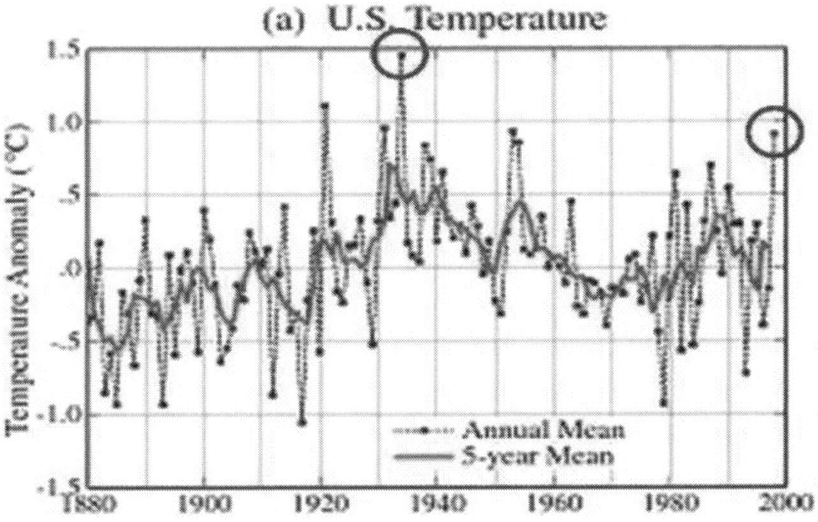

The Revised chart below with *"necessary revisions for anomalies,"* shows the warmest years to be in 1998, 2007. (Then, they contend, the current years are the warmest yet.)

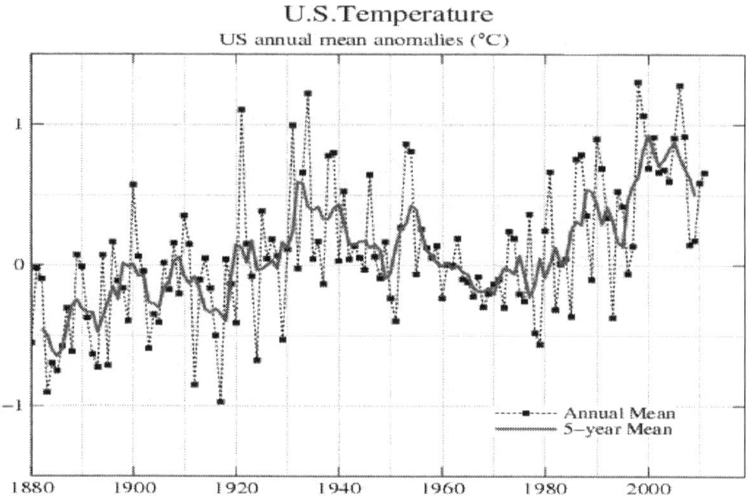

Even conceding their revisionist history, their claim of "warmest ever," only covers the last 140 years. What they fail to mention, is that the current "warmest ever" temperatures are, in fact, still part of one of the coldest periods of the past 600 Million years. (Refer back to the chart on Page 10.)

So, *ignoring* the argument of what is causing the current warming for the moment, any reasonable person would agree that the warming is beneficial, since it is bringing us out of the most recent Ice Age and one of the coldest periods in Earth's history. I fail to understand the Alarmists concern for warming of **0.**7°C in the last century. AND their alarm for a predicted warming of 2.5°C to 6°C is ***based solely on projections of Global Climate Models (GCMs) which have a history of inaccuracy, always predicting much more warming than actual results.***

<u>The truth of the matter...</u> ***This whole thing really isn't a story about Climate Change***. It's a story about stopping industry in wealthy, industrialized nations and bringing about a "One-World Government."

If you wanted to make the engine of an automobile (or that of an industrial plant and even an industrial nation) stop running, you could either stop the flow of fuel into the engine or stop exhaust from coming out. Either one will make the engine quit running. The Alarmists figured out early on that they couldn't likely stop fuel supplies worldwide, BUT, they could plug up the exhaust.

"***How could they do that?***" You ask. By scaring the populace into believing that the exhaust from our automobiles and our industrial plants is destroying the planet, and will soon bring about Armageddon.

They needed to paint a picture so dire, so desperate, that they could then say, "No nation on its own can solve this overwhelming problem. We must have a One-World Government to bring about, and then enforce, the necessary changes to save our planet."

So, like a magician using distraction to accomplish the "magic" of his performance, the Alarmists use the debate over Global Warming as their distraction.

The never ending debate keeps everyone's focus on just that... The Debate. Meanwhile, they continue to ramp up the scare tactics hoping to strike enough fear into enough of the populace that a majority finally gives in to their solution. The solution being a One-World Government ("It's the only way to accomplish the changes necessary to save our planet," they say, "since no one nation could do this on its own.")

And along with that, of course, comes a heavy tax burden on the wealthy, industrialized nations of the world.

The good news, in my opinion, is that the Alarmists vastly underestimate the resoluteness of the USA, and the stubbornness and inflexibility of Russia and China. Neither China nor Russia would agree to global governance such as this, unless they were the ones in charge of it all. And we in the United States would never give up our sovereignty and freedom (or the comforts of a wealthy, industrialized nation).

So, thankfully, I don't think it will ever happen. (Though the Alarmists don't seem to agree and continue with more and more "alarming" stories to tell.)

Chapter Five

Now (against all sensibility) we return to the debate over the cause of the recent warming. Actually, I have a difficult time imagining the level of narcissism, or the size one's ego, that's necessary to think you can have a significant influence on climate. Earth has survived for 4.5 billion years, taking everything the Cosmos could throw at it. It will certainly survive comfortably while humans burn fossil fuels.

I have recently been reading a lot from Russian scientists. They first got my attention when one of their Climate Models was the nearest of all 100+ models to the actual temperature results for the past two decades.

It's only my opinion, but the accurateness of their models and their science comes from their scientists having no need to accommodate politics and politicians. The Russian government (Putin et al) wants/demands "truth" from their scientists, not political manipulation.

One such scientist is Habibullo Abdussamatov, a Russian Astrophysicist known for his work with the physics of the sun. He is Supervisor of the Russian section of the International Space Station and is definitely a skeptic of "Man-Made Global Warming." He contends global warming is caused by the natural processes of the Earth and the sun. He considers himself to be among the many scientists who dispute the claim that, *"The science is settled."*

He further contends that global warming has been the result of an unusually high level of solar radiation, along with a lengthy – almost all of the last century – growth in its intensity. "The parallel, concurrent global warmings that have been observed on Mars and on Earth can only be a consequence of the effect of the same factor: the long-term change in solar irradiance."

Abdussamatov also contends that Earth's atmosphere does not operate with a greenhouse-like effect. "Ascribing 'greenhouse' properties to the Earth's atmosphere is not scientifically substantiated." The following diagram is a representation of some of the systems affecting our atmosphere. It demonstrates how complex it is and how difficult it is to reduce it to a computer program.

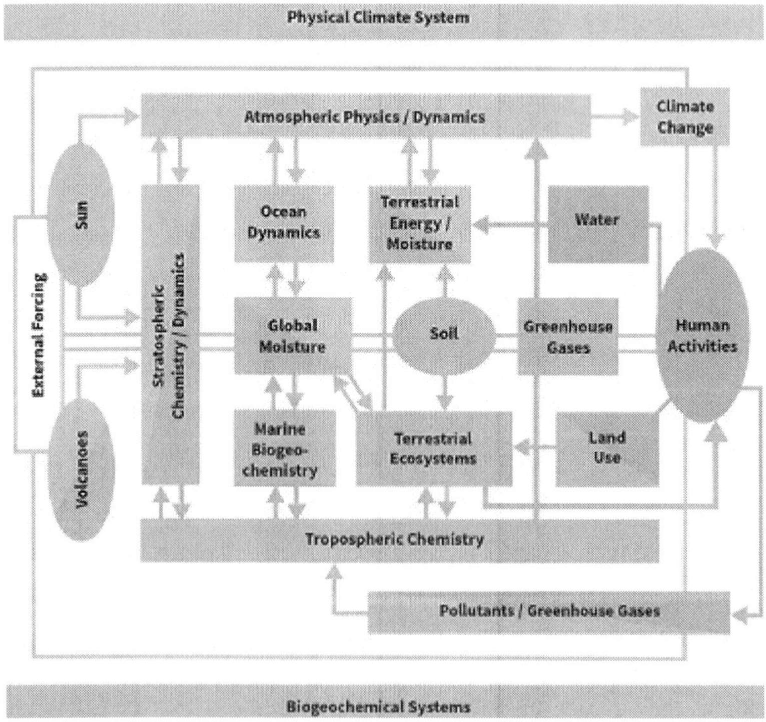

The Alarmists (those who are continually bolstering their Man-Made Global Warming theory) have solved it. They've analyzed and evaluated and narrowed the whole complicated system down to one minor participant that they can regulate and control – CO2. And narrowing it further still just the 3% of CO2 emissions from human activities while ignoring the 97% from natural causes. That's the Control Knob for Earth's temperature. That's the thermostat.

The diagram below depicts Earth's total atmosphere compared to CO2 emissions, and the 3% caused by humans.

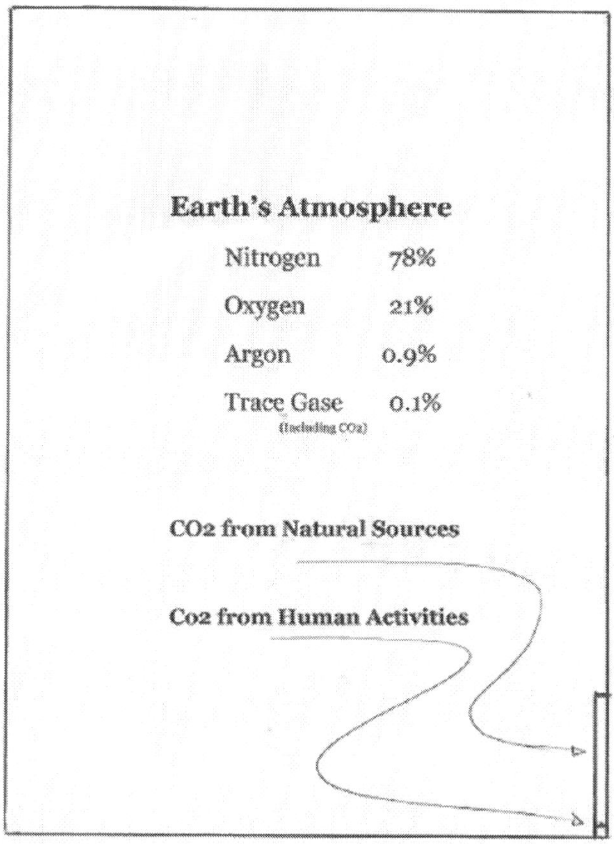

The diagram on Page 19 demonstrates the immense size of our atmosphere compared to the relatively insignificant amount of CO_2 emitted every year by human activities. Still the alarmists contend that by controlling this minuscule portion of our atmosphere they have the ultimate "Control Knob" for global warming.

Chapter Six

The Alarmist Thesis: CO2 causes global warming.

The only place in the history of scientific records where this happens is in their Global Climate Models. In reality, in every historical climate record, warming occurs first, then the CO2 level rises. In Earth's history, the rise in CO2 generally lags behind warming by a few hundred years.

Speaking of GCMs, let's look back at Moscow. In a Science Matters paper, "Temperature According to Climate Models," only one of the models was accurate in predicting temperature fluctuations.

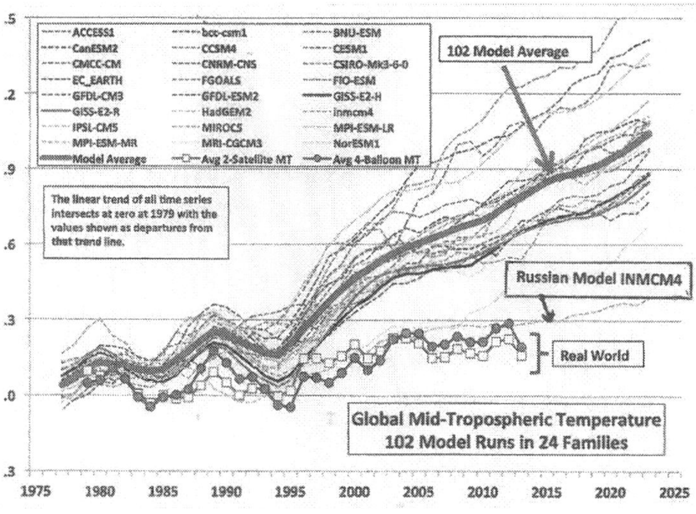

And, the most accurate of the 100+ models is (drum roll please), the Russian model designated INMCM4. It also projected an un-alarming 1.4°C warming out to the end of this century. In contrast, the other models have all predicted excessive warming over the past two decades and forecast future warming of as much as four times the 1.4°C of the Russian model for the end of the 21st Century.

The Page 21 graph shows the results from 102 GCMs. The Russian INMCM4 is the light dashed line running close to the "real world" results. You can easily see the bold line, the *average* of the 102 models. And, you can see that the *average* predicts considerably more warming than actual recorded temperatures. That's a result of having too much forcing for increased CO2 that's programmed into the models. The warming that they contend will be Catastrophic is based fully and completely on the 100+ GCMs that have a long history of predicting too much warming.

The INMCM4 was programmed by modelers at the Institute of Numerical Mathematics of the Russian Academy of Sciences, Moscow, Russia. Again, the Russian science and Russian scientists seem to be much more accurate with their work, their models, and their projections. It's unfortunate that so many of our scientists (and those of other countries as well) are so heavily influenced by the huge government grants available to those who are endorsing the "right" theories. Real science can easily be thrown aside when there's millions of dollars at stake.

That brings us to the deplorable acts of "Mann's Hockey Stick," "Data Doctoring," "Climategate," and the fraudulent, shameful rewriting of historic climate records to relentlessly prop up their thesis.

Dr. James Hansen... "The Climate Chiropractor." Need some temperatures adjusted? Call Dr. James Hansen. (You can do so much cool stuff when you have control of the historical records.) BUT, in 1999, Dr. James Hansen made these interesting, remarkable comments:

"Empirical evidence does not lend much support to the notion that climate is headed swiftly toward more extreme heat." ***(aka There's nothing alarming about future climate.)***

"At a time of rapidly increasing CO2, in the U.S. there has been little temperature change in the past 80 years – in fact there was a very slight cooling trend throughout most of the 20th century." ***(aka Cooling trend... Not catastrophic warming.)***

GISS US temperature data showed an 80+ year decline in US temperature averages, with 1934, 1921 and 1931 being the three hottest years.

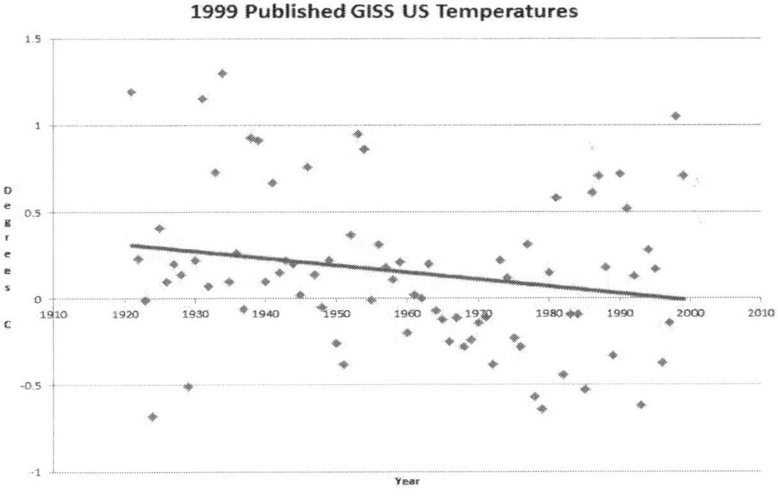

(Hansen's comments and graph have been deleted from the GISS web site, but originally did reside there.)

This next graph was published in 1995's Second Assessment Report of the Intergovernmental Panel on Climate Change (IPCC), and has been around for decades.

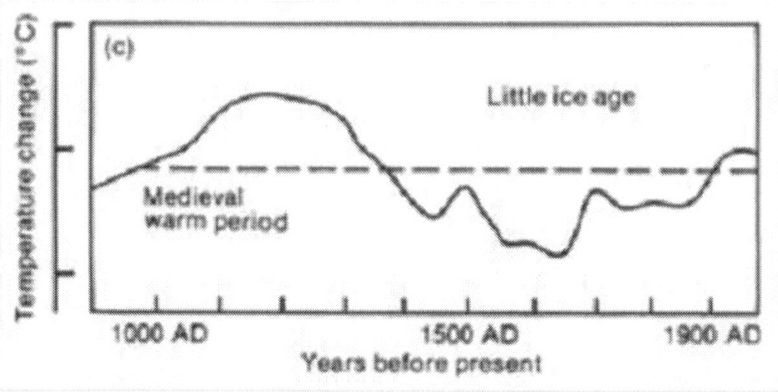

The graph clearly shows both the Medieval Warm Period and Little Ice Age. These have been well established with geological as well as antidotal evidence. There have been nearly 2,000 papers written on these two historic periods in our climate.

In 2001, in their Third Assessment Report, the IPCC removed that chart and featured "Michael Mann's Hockey Stick graph," using it five or six times in their report and wiping away the Medieval Warm Period and Little Ice Age. Mann's "reconstruction" of temperatures for the past 1,000 years was accomplished by manipulation of tree ring studies until about 1950.

When those same tree ring studies showed a decline in temperature as he continued forward, he instead spliced on instrumental temperature readings from 1950 to the present. Even a non-professional, like me, would question the efficacy of stitching together two different source types to achieve a "scientific" result. Graphs similar to the one shown on Page 25 were used several times in the 2001 IPCC report.

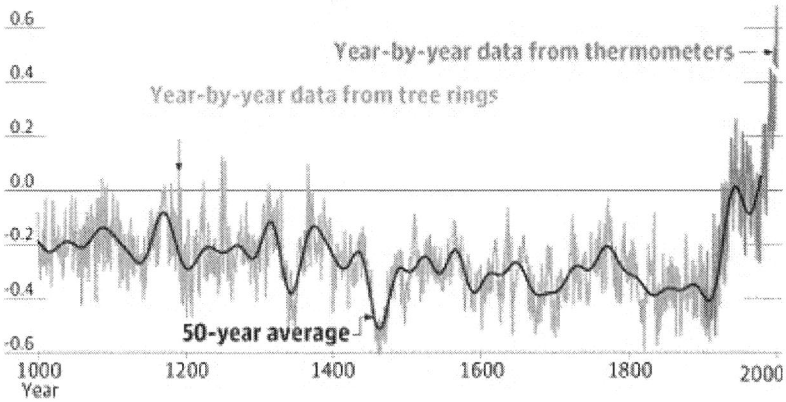

Stephen McIntrye and Ross McKitrick evaluated the data used in the Mann paper and concluded that the Mann curve was invalid, "Due to collation errors, unjustifiable truncation and extrapolation of source data, obsolete data, geographical location errors, incorrect calculation of the principal components, and other quality control defects." Since then, information disclosed in the Climategate scandal has revealed how the data was manipulated.

Mark Steyn, one of several individuals that Mann has filed law suits against, said of Mann's proxy reconstruction resulting in the hockey stick graph, "There are only two things wrong... The proxy and the reconstruction."

This brings us to the next issue: "Climategate." It began in 2009 when a whistle-blower leaked about 1,000 emails. The emails were between a few scientists central to the "Human caused global warming" crisis. Then, two years later, a new batch of nearly 5,000 emails between those same scientists were released to the public, igniting another firestorm. In the emails they discussed:

1) *Efforts to exclude "skeptics and deniers" from publishing in journals.*

2) *Punishing editors who allowed any denier articles to appear in their journal.*

3) Blocking requests for original data and, when necessary, manipulating the data used to fit desired outcomes.

5) Rush into publication all articles that refute or attempt to discredit scientists who disagreed with IPCC's position."

Below is a copy of an email sent on November 16, 1999. It was sent by Keith Briffa, Climatologist, to three of his fellow Alarmists, and refers to "Mike's Nature trick..."

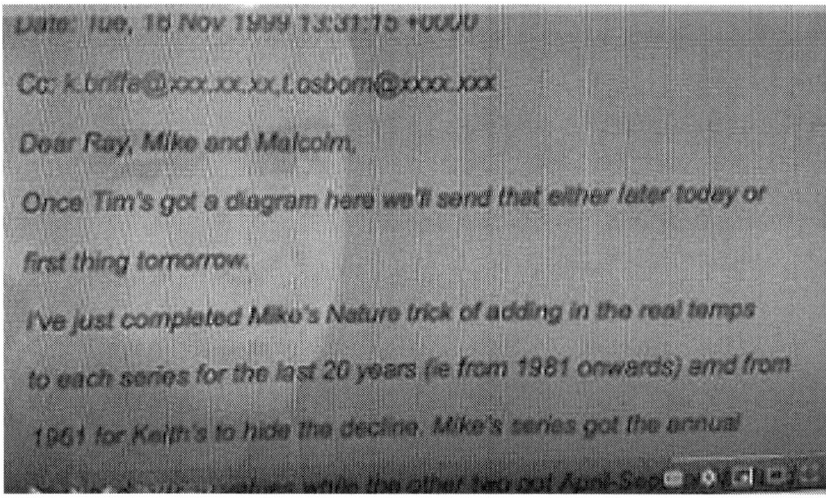

Date: Tue, 16 Nov 1999 13:31:15 +0000

Cc: k.briffa@xxx.xx.xx,t.osborn@xxxx.xxx

Dear Ray, Mike and Malcolm,

Once Tim's got a diagram here we'll send that either later today or first thing tomorrow.

I've just completed Mike's Nature trick of adding in the real temps to each series for the last 20 years (ie from 1981 onwards) amd from 1961 for Keith's to hide the decline. Mike's series got the annual

Three themes emerged from the released emails:

(1) Prominent scientists were taking measures to conceal rather than disseminate the underlying data.

(2) Many of these scientists view man-made global warming as a ***cause*** rather than a scientific inquiry.

> Michael Mann wrote the following: "I gave up on her (Climate Professor, Judith Curry, Georgia Institute of Technology) a while ago. I don't know what she thinks she's doing, but it's not helping *the cause*."

(3) Many of the emails admit *their science is weak and dependent on deliberate manipulation of data.*

Tom Wigley, scientist at the University Corporation for Atmospheric Research wrote this to Mann: "Mike, the Figure you sent is excellent... Very deceptive. <u>There have been more than a few dishonest model results by authors and by the IPCC."</u>

"Man-Made Global Warming" would indeed be a valid concern of ultimate global importance (If it were it true). It would be a concern that folks would have thought *should demand the most careful, exacting and rigorous scientific analyses that mankind could muster.*

Instead of seeing a large collaboration of meticulous, careful, critical scientists, *Climategate* gave us a peephole into the work of a small group of "scientists" building a wall around them to prevent any real scientists from seeing their questionable, sometimes fraudulent research data. *Global Warming* became a *Movement* in search of the *Huge Funding* available from governments around the world.

• • • •

The latest bit of alarmism was an article saying, **"We are losing the battle to the rising seas and must cut our CO2 emissions <u>*immediately*</u>."**

Are the sea levels continuing to rise? Yes, they are, but at roughly the same 2.2 mm per year, or about 11 inches per century, since we began keeping records two hundred years ago.

The following graph had the headline: "Why is the relative sea-level in Hudson Bay <u>*falling*</u> so much?

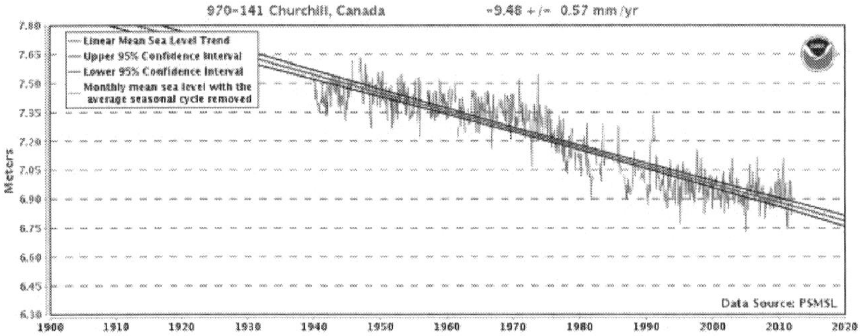

This graph charts the average sea-levels for Churchill, Canada on Hudson Bay, and shows a lowering of sea-levels of about 0.95 meters (roughly 3 feet) over the last 100 years. Hudson Bay is connected to the world's oceans, primarily through the Hudson Strait, so how can this be? The reason... The area has been experiencing "post-glacial rebound." The land is rising.

During an Ice Age, the extraordinary weight of glacial ice sheets actually depressed the Earth's crust creating huge swales or "dents." This lowering of the surface level can take a protracted amount of time to 'bounce' back.

All over our planet land levels are rising and falling (at a relatively slow rate of course). Some, such as this Hudson Bay example, experience post-glacial rebounding; some from slow continental drift; some from plate tectonics; and some for other reasons.

What's the point of all this? The next time someone wants to tell you sea level is rising at a high rate in some particular place, ask them, "What is the land doing?" Chances are they'll have no idea.

Chapter Seven

I would think that if one set out to champion an "alarming," "catastrophic," "moral" cause such as the tumultuous "Man-Made Global Warming" (the original label), that one would first have an abundance of scientific evidence in one's favor. Instead, they have been continually caught with their "hand in the cookie jar" manipulating data to prop up their argument at every turn. They surely must recognize that they're fighting a losing battle. And they surely must know the devastation they've brought to the once solid reputation of science – not just climate science, but science in general.

Nearly three decades later, we still see evidence of data doctoring to prop up their theory (their lost "cause"). I would be embarrassed by the fact that it's become too obvious, even to the general public, with the constant stream of "Necessary Adjustments," "Required Refinements," "Homogenizations," "Anomalies," "Michael Mann's Hockey Stick Graph" (accomplishing what was requested by more than a few leading Alarmists, that "We've got to get rid of that Medieval Warm Period"). WHY? How can they claim this warming to be "unprecedented" with a period warmer than now only 160 years ago? And no carbon footprint on which to blame that warming. They had to get rid of it.

The graphs on the next page depict some of the data doctoring and manipulating of historical records that's been done to climate records all over the world.

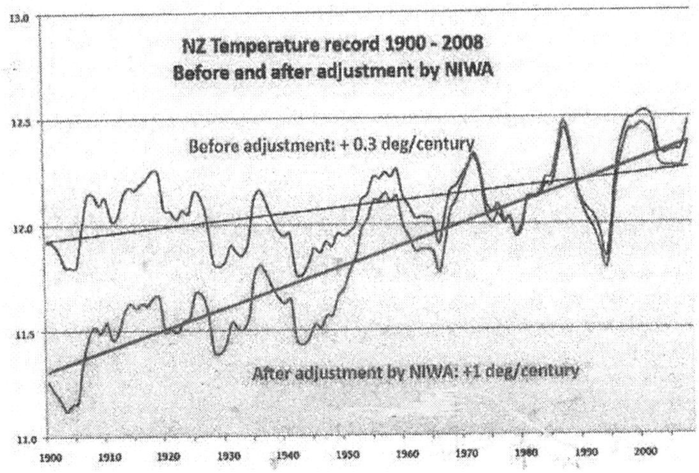

The above graph is from New Zealand's National Institute of Water and Atmospheric Research (NIWA), comparing "before" and "after" the adjustments were made to historic records back to 1900. (Earlier temperature records were lowered to exaggerate the warming trend.) Much the same was done by Australia's Australian Bureau of Meteorology, as well as many other government agencies around the world. The following graph shows similar results from needless "Homogenization" of temperature records.

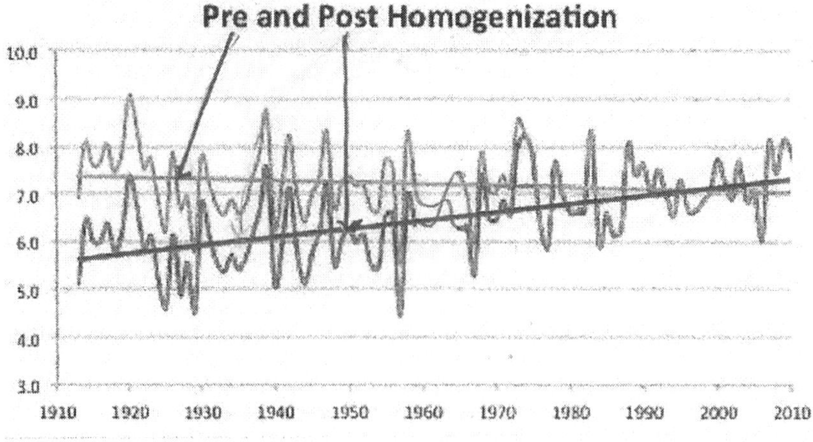

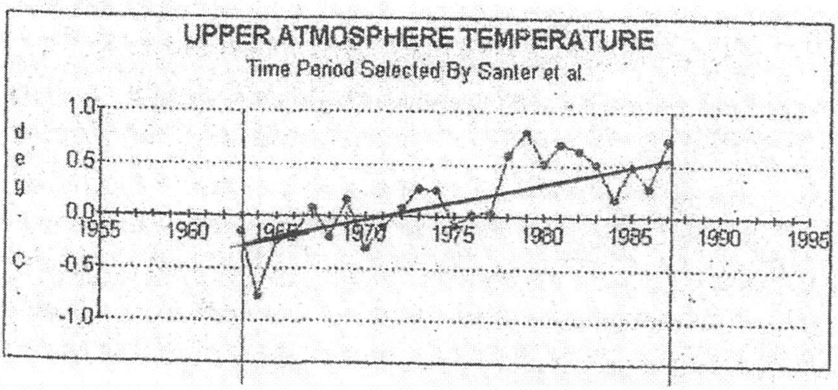

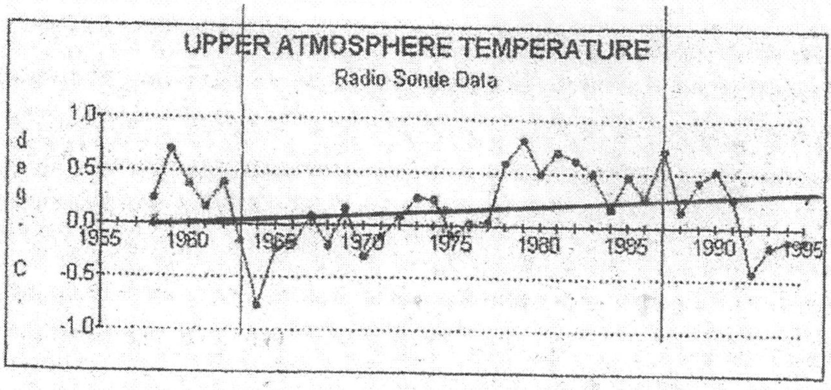

The above charts are examples of "Cherry Picking." To influence trends on a graph, you can manipulate the results by picking the endpoints. In the graphs above, the designer of the first chose arbitrary ends at 1963 and 1987 which showed a strong warming trend. But on the second, using the full scale of 1955 to 1995, it shows a very slight warming trend (both graphs are based on the same data).

There are a number of “tricks” that can influence the data to tell a story that supports your argument. Cherry Picking is one. Manipulating the scales of the X and Y axis can show you a very different picture, as in the two charts that follow.

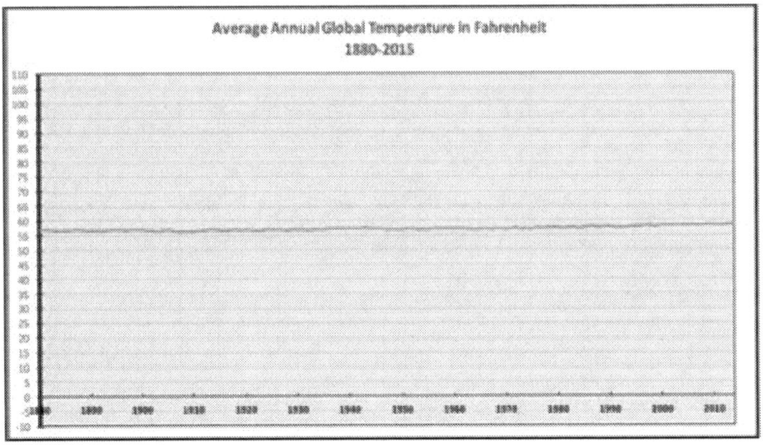

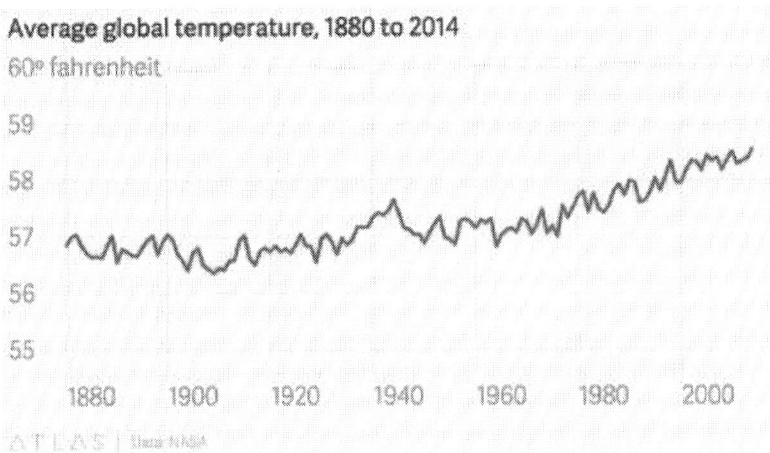

Using the same data, but changing the scales for the vertical and horizontal axis, we get very different pictures of the average global temperature from 1880 to 2015.

It's amazing that EVERY form of "Data Doctoring" results in strengthening the Alarmist's argument.

Chapter Eight

How many more methods of perpetrating the fraud that is "Man-Made Global Warming" can the Alarmists come up with? "Let me count the ways," (to quote Browning). In my first book on climate, "A Brief Look at Earth's Beautiful, Chaotic Climate," I had fifteen chapters detailing the many facets of this colossal fraud.

As Dr. Timothy Ball, retired Professor from the University of Manitoba, said, "This is not a hoax. A hoax is malicious, but there's generally something humorous behind it. There is nothing humorous about the Alarmists theory of 'Man-Made Global Warming,' especially to the poorest of the world's poor nations. No, this is the biggest ***deception*** ever perpetrated on mankind."

Why would Al "Oscar Award Winning" Gore ask we the people to "... set aside water vapor for the moment" in the discussion of greenhouse gases in his Academy Award winning fraud... (oops... movie), An Inconvenient Truth? Could it be that including water vapor didn't serve his purpose since it has an immense effect on climate (about 96% of the effect from greenhouse gases) whereas CO2 has a miniscule effect by comparison?

The concentration of water vapor in the atmosphere is approximately 50,000 ppm, while the current level of CO2 is approximately 400 ppm.

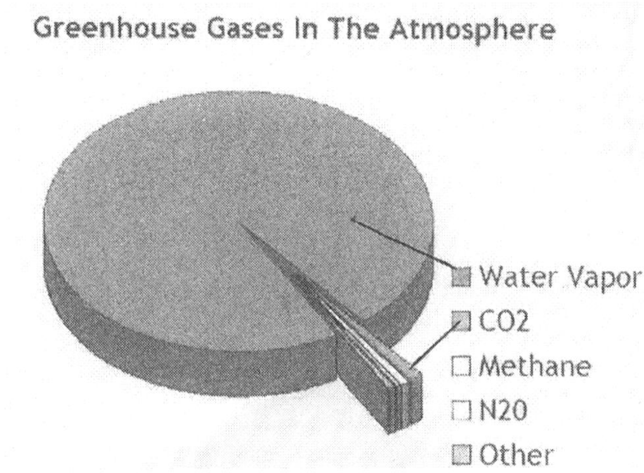

The Alarmist's theory on the influence of water vapor is a good example of circular thinking. It says, "As the climate warms due to human-caused increases in CO_2 (Anthropogenic warming), water vapor increases and traps more heat in the atmosphere causing more warming."

First, you have to accept their theory of Anthropogenic (Man-Made) Global Warming as already proven. (The problem is that it's a theory for which there is ***absolutely no scientific proof***.) So, you first must accept their flawed theory; that then will lead you to the true assumption that this will cause more water vapor, which will cause more warming. But it's a circle (or circular thinking) based on a flawed premise.

Next, let's look at the Beer-Lambert Law. The equation representing the law is straightforward: **A=ebc** Where A is absorbance (since $A = log_{10} P_0 / P$) **e** is the absorptivity with log10 standard linear units of L $mol^{-1} cm^{-1}$ **b** is the path length **c** is the concentration of the compound in L-1...

(Eyes glazing over.)

Actually, this Law is only an approximation in terms of CO_2 and its radiative forcing causing warming, because water vapor contributes about 96% of the "greenhouse" warming. In "plain English" it says that CO_2's effect on warming continues to rise until you reach a certain level (at our current 400 ppm we are beyond that level) then there is very little additional effect when adding more CO_2.

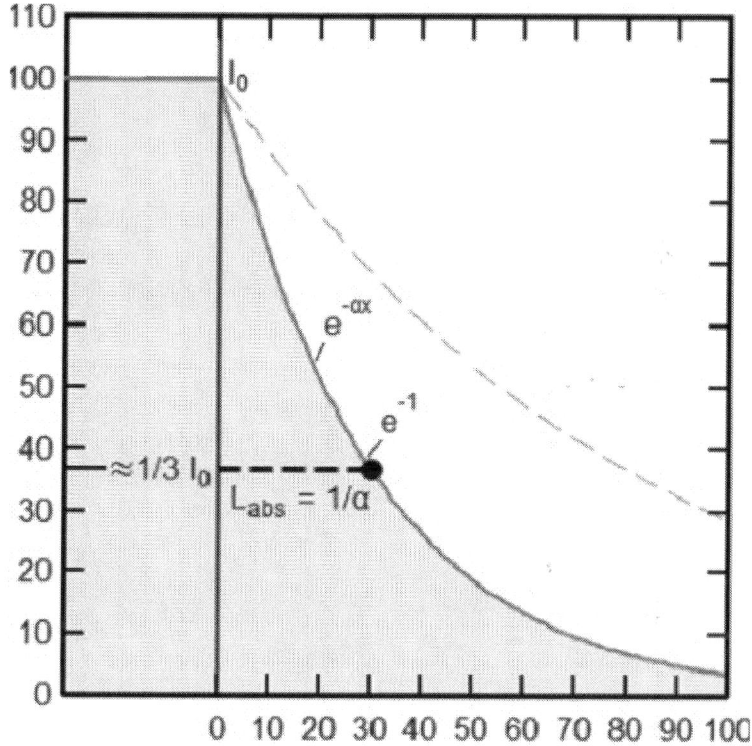

A friend explained it this way. If your barn was green and you want it red, the first coat may not quite cover all the green. With a second coat, your barn would likely be a nice red. Painting it with another coat would do very little to make it "more" red. Adding more CO_2 to our atmosphere today, would be like adding a third or fourth coat of red paint to your barn. The effect would be almost immeasurable.

The Alarmists contend that doubling the CO_2 content would raise atmospheric temperature 1°C. So, doubling our CO_2 level from 400 ppm to 800 ppm they say would raise Earth's temperature by the 1°C. Again, doubling from 800 ppm to 1600 ppm would raise another 1°C. (Others question the 1°C with each doubling of CO_2, but we'll ignore that argument for now.)

Since about 1950 when human activities increased significantly, the industrial age kicked into high gear, and CO_2 emissions grew (remember that the human caused emissions are only 3% of the total CO_2 emitted annually) *the level of atmospheric CO_2 has only increased 100 ppm – from 300 to 400.* Assuming a steady increase in human activities world-wide, it would still take two centuries, maybe more, to double from 400 ppm to 800 ppm. A warming of 1°C in two centuries would be beneficial for most of the world, not catastrophic.

Chapter Nine

The Alarmists treat the Sun a “constant” in relation to climate. They treat it that way in their theory and that way in their climate model programming. The sun is definitely not a constant, and many scientists (again a Consensus?) consider it the biggest factor in Earth’s climate.

Throughout the Earth’s five major Ice Ages there have been numerous glacial periods happening in roughly 40,000 and 100,000 year cycles, with warm interglacial periods between. It is thought that these glacial and interglacial periods are largely due to the variations in Earth's orbit and the tilt of its axis.

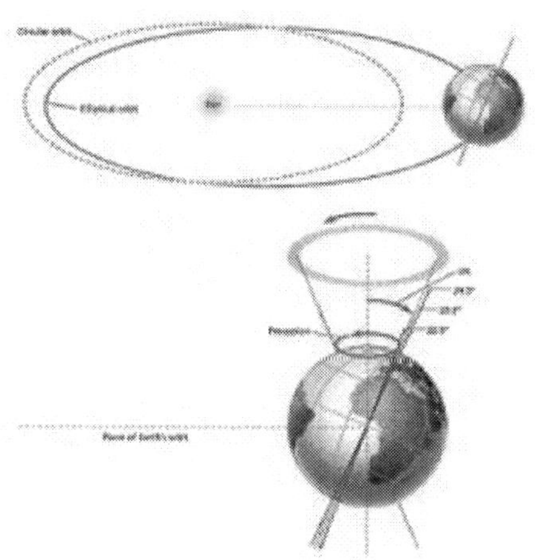

Milankovitch’s Orbital Variations
Axial Tilt, and Precession

Serbian mathematician, Milutin Milankovitch, determined that glacial cycles were influenced by Earth's orbital variations, eccentricity, and procession. These all affect solar radiation levels that reach the Earth's surface and contribute more energy/less energy to warming.

The "Milankovitch Cycles" are defined by these three major variations:

> *__Eccentricity__ - The changes in Earth's orbit cycle over a 100,000 year period. It slowly changes from a near circular orbit to an elliptical one, then back. This is from the gravitational influence of the other planets, particularly Jupiter.
>
> *__Obliquity__ - The tilt of the Earth's axis ranges from 22° to 24.5°, and transitions over roughly a 40,000 year cycle.
>
> *__Precession__ - The wobbling movement of the Earth's axis as it rotates. The axis rotates in a circular pattern over a 26,000 year period.

The action of all three of these cycles can create situations where, when everything is lined up just right, the sun will provide the least amount of radiation to Earth. This would cause cooling and maybe even an ice age. The opposite would also be true with warming.

Obviously, the Milankovitch Cycles are not the only factor affecting our climate. Others would include oceans and ocean currents, water vapor, clouds, precipitation, Southern Oscillations, South Pacific Convergence, North Atlantic Oscillations, El Niño, Intertropical Convergence, the Hadley Circulation, the Madden-Julian Oscillations, atmospheric blocking, tropical cyclones, storm tracks, jet streams, and more. (But the Disciples of Man-Made Global Warming keep telling us, "Ignore all that... It's only our CO2 emissions.)

Chapter Ten

At more than 2,000 pages, few scientists have read the full reports of the IPCC (and of those that have, few understand fully what it says). Most affiliated scientists are only working on one small portion of the report. They trust that those working on other areas of the report are being truthful and are uncorrupted.

More than a few scientists have defected from the ranks of the IPCC. Klaus-Ekert Puls, a German meteorologist and physicists, was shocked at what he found after digging deeper into the details of what the IPCC's reports proclaimed. He wrote, *"For more than 10 years I simply parroted what the IPCC told us. When I started checking the facts and data, I did so with a sense of doubt, but then became outraged when I discovered that much of what the IPCC and the media were telling the world was sheer nonsense, and* <u>*not supported by any scientific facts or measurements.*</u> *To this day I still feel shame that, as a scientist, I had made presentations of their "science" without first checking it myself."*

Dr. Mototaka Nakamura received his Doctorate of Science from the Massachusetts Institute of Technology (MIT), and for nearly 25 years specialized in abnormal weather and climate change at prestigious institutions that included MIT. After breaking ranks with the IPCC, he said, "Their models are simply mockeries of the real world."

He explained that the data underpinning the Global Warmist's science is doctored, untrustworthy, and cannot be verified. "Merely questioning or testing the validity of their claims is regarded as ***heresy*** (again a religious orthodoxy). And now there being fronted by frightened, crying children and teens (see Greta Thunberg) who are completely uninformed and misled.

Nakamura says that data falsification is prevalent throughout those working in global warming and that when arguing against it, the hardest thing is to convince people of is the data falsification and temperature doctoring. "If you don't pick your words carefully, or get your tone exactly right, then it's very easy to sound like a conspiracy crank."

He goes on to say that their climate models grossly simplify the way climate works. As well as ignoring the Sun's effect, they drastically simplify ocean dynamics, water vapor, and clouds. "Earth's climate system is far more complex than the absurdly simple systems that are simulated by their climate models. And solar input is modeled as a 'never changing quantity,' which is absurd."

A very interesting item which is almost unreported...

The Recent Warming on the planet Mars.

Global Warming is happening on Mars, as well as Earth. A study of its ice caps suggests Mars is experiencing a warming trend. Data from NASA cites a six year study by researchers at Duke University_(as well as Russian scientists mentioned on Page 18) shows Mars is seeing a warming trend, and that both Earth and Mars are seeing simultaneous warming resulting from the sun's output.

Planetary scientists have been watching the melting of deep, wide pits in Mars's southern ice cap. The recent melting is substantial. According to Michael Malin, principal investigator for the Mars Orbiter Camera, the polar ice cap is shrinking at "a prodigious rate."

The most recent images from the NASA study show changes during the early 2000s suggest the climate on Mars is presently warmer, and perhaps getting warmer still, than it was a couple decades ago.

If both Mars and Earth are experiencing global warming, then it's logical there is a larger phenomenon going on in the Solar System that is causing their climates to change similarly – like changes in the level of the Sun's radiation perhaps. (Or do the Alarmists actually think that our "Anthropogenic Global Warming" is somehow contagious or rubbing off on Mars?)

Part Two

– The Cooling –

Grand Solar Minimum

Notice to All Acolytes of Catastrophic Global Warming:

After many studies – and an in-depth research project by some of the top minds available in climate catastrophe studies – it has been determined there is no way to tax individuals for the Sun. Therefore, the Sun is *ruled out* as a cause for climate change.

Chapter Eleven

The sun is a huge ball of hot, fluid, electrically charged plasma, that's 1.3 million times the size of Earth. It is the most important source of energy for all life on Earth, and is continually moving, rotating, turning, and generating a powerful magnetic field.

Sunspots are temporary phenomena on the sun's surface that appear (to observers on Earth) darker than the surrounding areas. They appear darker because they are areas of reduced surface temperature. The temperature of a sunspot is approximately 7,000°F, compared to the sun's surface temperature of 10,000°F. Sunspots have been observed since the ancient Chinese first made reference to them centuries ago, and they have been studied by scientists since the invention of the telescope in about 1610. The "solar cycle" was discovered in 1843 by S.H. Schwabe after two decades of studying the sun and sunspots.

The sun exhibits cycles lasting approximately 11 years. The cycles are marked by the increase and decrease of sunspots, the visible dark blemishes on the sun's surface. The highest point in the cycle (with the greatest number of sunspots) is called the "solar maximum." The lowest point or lowest number of sunspots is the "solar minimum." These cycles can vary from 9.5 to 12.5 years, but over 170 years they have averaged 11 years per cycle.

The magnetic field lines near sunspots often get tangled, crossed, and reorganized. This can cause a sudden explosion of energy called a *solar flare*. Solar flares release huge amounts of radiation into space. If a solar flare is intense, the radiation it releases can interfere with radio communications here on Earth. Solar flares are sometimes accompanied by a coronal mass ejection (CME). CMEs are huge bubbles of radiation and particles of the sun that will explode into space at extremely high speeds when the Sun's magnetic field lines suddenly reorganize.

Levels of solar radiation and ejection of solar material, the number and size of sunspots, solar flares, and coronal loops all exhibit a similar fluctuation, from quiet to active to quiet again in the same 11 year cycle. The cycles have been observed for nearly two centuries, but no one knows what causes them or what determines the length of them. Besides the amount of sunspot activity rising and falling in the 11 year cycle, the sun's magnetic field reverses polarity with each cycle – the north pole switching to the south in one cycle, and back again in the next.

Individual sunspots or groups of sunspots may last anywhere from a few days to a few months, but they will eventually decay. Sunspots expand and contract as they move across the surface of the Sun, with diameters ranging from small (10 miles wide) to huge (100,000 miles wide).

Sunspots have two parts: the central umbra, which is the darkest part, where the magnetic field is approximately vertical, and the surrounding penumbra, which is lighter, and where the magnetic field is more inclined.

When charged particles from a CME reach Earth's atmosphere they can trigger intense lights in the sky called auroras. A particularly strong CME can also interfere with electric utility grids and cause power shortages and power outages. Solar flares and CMEs are the most powerful explosions in our solar system.

Scientists first began numbering the sun's cycles in 1755, with Solar Cycle 1 (SC 1). SC 24 just came to an end and we are in the solar minimum beginning stages of SC 25. We have been experiencing a steady decline in solar activity, with the last four cycles, SC 21 – SC 24, each having less activity than the previous (as shown in the following graph).

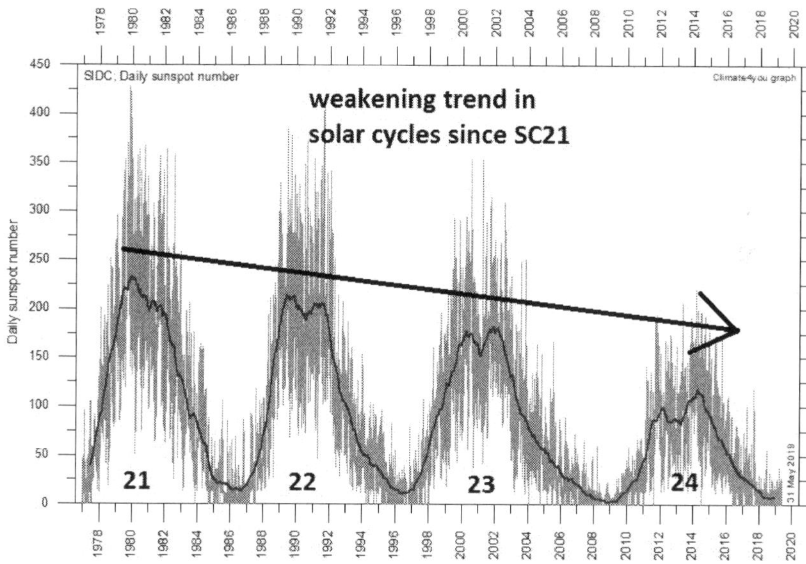

Several predictions have been made for Solar Cycle 25 ranging from moderately weak to very weak in magnitude. Most predict it will be the fifth consecutive cycle with diminishing activity. In June, 2019, NASA predicted, "The next solar cycle, SC 25, will be the lowest activity we've seen in 200 years. Dalton Minimum levels could be reached."

What is a **Grand Solar Minimum**? The next three or four solar cycles are projected to have lower solar activity, fewer sunspots, than the current one that's just ended. This indicates SC 25, 26 and 27 will be cycles with very little or no sunspot activity.

Extended periods of inactivity like this, is the definition of a Grand Solar Minimum. A GSM will likely cause global cooling and perhaps even a mini ice age.

The Little Ice Age (LIA) was dominated by four consecutive solar minimums: The Wolf, Sporer, Maunder, and Dalton solar minima. All of which lengthened the cold of the LIA to roughly 450 years.

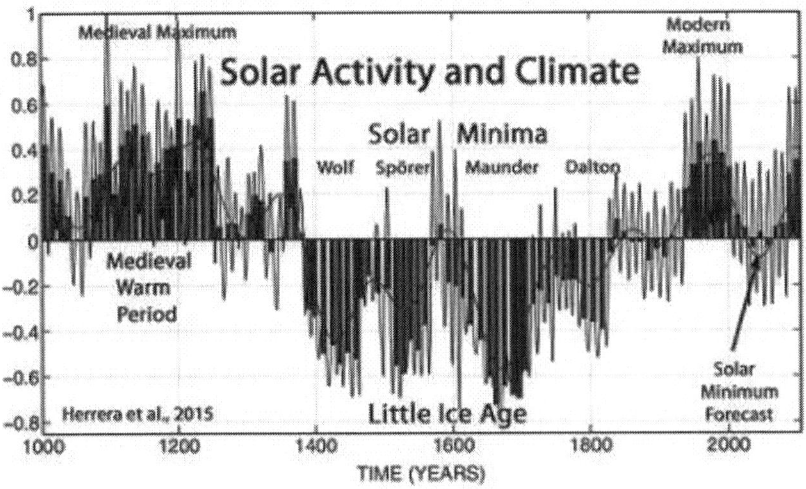

The Little Ice Age brought longer, colder winters to Europe and North America. With that came shorter summers and growing seasons, crop losses, food shortages, famine and death. In the 17th and 18th centuries, farms and villages in the Swiss Alps and Norway were destroyed by encroaching glaciers. Canals and rivers in Great Britain and the Netherlands were frozen, and early settlers in North America faced exceptionally severe winters.

Chapter Twelve

When there are lots of sunspot, solar flares and coronal mass ejections, the Sun releases much more radiation. This greatly adds warming to Earth's climate. With fewer sunspots, the Earth's climate grows cooler over time. Simply put:

High solar activity = Warmer climate on Earth.

Low solar activity = Cooler climate on Earth.

The following graph shows the solar activity from 1600 through 2020, and then graphs the predicted low solar activity from 2020 to the end of this century.

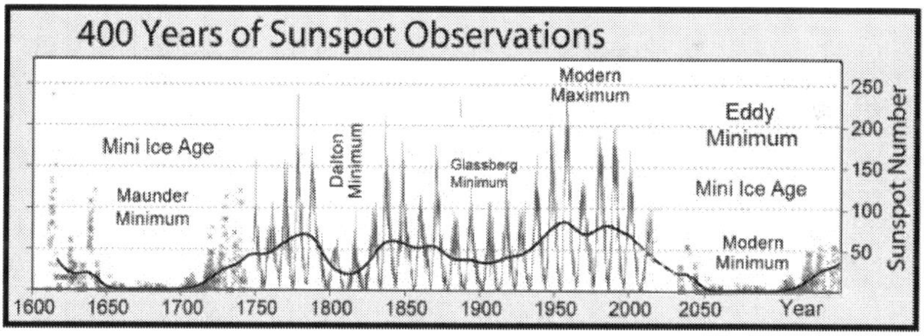

The graph shows the Maunder Minimum, which coincided with the Little Ice Age (on this graph it is referred to as the Mini Ice Age) and likely was the primary cause of it. The Solar Minimum we are now entering is called the "Eddy Minimum" named after the Solar Physicist John A. Eddy, who originally named the Maunder Minimum for British Astronomer Edward Walter Maunder.

If we are entering an extended Grand Solar Minimum that will last through three solar cycles, as predicted by many solar scientists, it means we will face longer, colder winters, shorter summers, shorter growing seasons, food shortages, and – in the poorer areas of the world – famine and death. It won't be a matter of just turning up the thermostat where you live to keep warm, it will mean finding enough food to feed your family, plus other hardships.

Earth's climate change continues (as it always has). However, there is strong evidence pointing to a totally different result than the IPCC and Acolytes have been proclaiming for the past three decades.

Earth's Climate is a huge, immensely complex system. So far, however, there are strong indications that the IPCC has ignored many important factors in our climate much to their/our peril. One particular factor they've ignored is the influence that sunspots and solar activities have on our climate. Rather than preparedness and readiness for a potentially devastatingly cold climate, we've been mired in a fruitless debate about why we had a brief warming.

The frequency and intensity of sunspot activity has been shown to have a profound influence on Earth's weather systems, atmospheric temperature, ocean temperature, Gulf Stream flows, and more. It is also notable that the IPCC simply dismisses these influences as insignificant.

Again, I turn to the Russian scientists... Professor Valentina Zharkova teaches mathematics at Northumbria University in England, and is a highly respected Solar Physicist who has studied the sun for two decades. At a conference of the Global Warming Policy Foundation in October, 2018, Zharkova presented her findings, saying, ***"In 2020 a Grand Solar Minimum period is expected to begin."*** (In the past, projections from her models have demonstrated 95% accuracy.)

Professor Zharkova also said, "During the Muander Minimum only two of the sun's magnetic fields went out of phase. This time all four fields will go out of phase." Evidence suggests that we need to develop a better understanding of the sun and the effect of solar activities to be able to prepare for major climate catastrophes in the future. A colder climate is far more devastating than a warmer one.

The United States Government's National Oceanic and Atmospheric Administration (NOAA) ***now tells us***, "All weather, all climate on Earth – from the surface of the planet out into space – begins with the Sun." Meanwhile, no climate model used by the IPCC or any of the hundreds of their climate think tanks around the world is able to accurately model the Sun. It is by far the greatest factor affecting our weather and they treat it as a simple constant.

It shouldn't be surprising that when confronting the possibility of a Grand Solar Minimum and global cooling, Alarmists respond with, "It will buy us time against the Catastrophic Warming that's coming." (Uugghh.)

In 2019, for more than three fourths of its 365 days there were no sunspots. It is the most sunspot-free days in any year since 1911. If we continue into a Grand Solar Minimum, it will be affecting cloud cover, the jet streams will be weakened, and we will likely see more earthquake and volcanic activity.

Dr. Mototaka Nakamura, climate scientist who defected from the IPCC in disgust, said, "Solar output is not a constant, as the IPCC insists. During solar minima, like the one we're entering, the sun's magnetic field weakens and the outward pressure of the normally strong solar winds decreases. This allows more Cosmic Rays (CRs) from deep space to penetrate our planet's atmosphere. Cosmic rays affect clouds, causing increased cloudiness – mainly low-level clouds that reflect the sun's energy back into space resulting in the cooling of Earth's climate."

He continued, saying, "There has always been an abundance of data to support natural variations in climate. Only the most narcissistic, those with enormous egos, would believe that human activities could have a significant influence on global climate."

• • •

H.L. Mencken said, "The whole aim of politics is to keep the populace alarmed – and hence clamorous to be led to safety – by menacing it with an endless series of hobgoblins, all of them imaginary."

The repeated wail of "Catastrophic Warming" is nothing more than one of the imaginary hobgoblins about which the insightful Mr. Mencken warned us.

• • •

And then there's the evidence of the last two decades. Since 1998 the atmospheric CO2 level has increased 20 – 25% (mostly from China, India and other "modernizing" countries), while the global temperature has remained statistically unchanged.

Even so, the Alarmists continue to announce "warmest ever," while dealing with hundredths of a degree. As Ivar Giaever, Norwegian Physicist with a Nobel Prize in Physics, said, "Do they seriously think they can arrive at an average global temperature for all of Earth's differing climates, express them in hundredths of a degree, and think that it's meaningful?"

Oh, by the way (in case you didn't know) Al Gore's mascot, the starving, dieing polar bear – because we're causing the Artic ice to melt – is doing just fine (which the lieing, deceitful Uncle Al, knew all along).

In the 1960s the polar bear count was 5,000 to 6,000 bears. That count, nearly 50 years later in the early 2000s (while Gore was producing his movie), was 25,000 to 30,000 – the increase mostly due to hunting restrictions.

In Conclusion...

Climate is changing. Always has... Always will. Earth has been through cooling cycles, warming cycles, Ice Ages, glacial and interglacial periods, and extended warm periods with no ice anywhere on our planet. All of these took place ***before there was any such thing as CO2 emissions from human activity***. Most of it even before Homo Sapiens were walking upright and appreciating this beautiful planet.

Therefore, the real causes, ***the natural causes for climate change***, we're at work bringing about warming and cooling (even the extremes that brought severe Ice Ages, then melted away the ice and warmed Earth) long before we were here.

The Alarmists are taking a page from Joseph Goebbels, Adolf Hitler's minister of propaganda, who said, "Repeat a lie often enough and it will become the truth." They have been beating the drum and repeating their "theory" for three decades. Fortunately, they've not been able to force truth and reality from the argument.

Fossil fuels have given us wonderful things... Longevity, prosperity, and personal freedom. And tomorrow these same benefits should be available to the developing nations of the world as well.

• • •

I've lived on this beautiful Earth for 74 years and have experienced all of the following ***scares***:

- That Earth's population is exploding and will cause world-wide famine. (The original goal and goal of all of the "scares" since – extreme population reduction.)

- The use of chemicals would cause cancer epidemics.
- That we were going to die from DDT.
- We were in fear of imminent nuclear holocaust.
- That we would soon run out of oil.
- That we were going to run out of fresh water for drinking.
- That we were on the verge of the next Ice Age.
- That we were going to suffer epidemic skin cancer and eye damage due to our depleting the ozone.
- In 1982 the United Nations Environment Program informed the world that we were at a tipping point of climate disaster and must take immediate action.
- In 1989, they issued another 10 year tipping point.
- A year later, in 1990, the UN's top official declared, "The world faces an ecological disaster in the next two decades unless governments act now."
- Then there was the millennium, when we were told computer systems would crash, utilities collapse, and airplanes might drop out of the sky. Our move from 1999 to 2000 was a big yawn.
- In 2009, James Hansen, then the director of the NASA Goddard Institute for Space Studies, said *"The President has four years to save planet Earth."*

None, I repeat, NONE of these actually came out as predicted.

Now these same "environmentalists" are telling us we're going to fry and die from man-made global warming. As mentioned before, the debate/argument is not really about CO2 emissions from human activity, or even about climate. It's about the same goals since the beginning: Crippling the "rich" industrial nations, bringing about a One-World Government, and drastically reducing in the world's population.

Prediction: The next 'Environmental Catastrophe' will be about an Extreme Water Crisis.

Made in the USA
Monee, IL
15 July 2020

35762732R00035